TORNADOES

ANNE SCHRAFF

Artesian Press
P.O. Box 355, Buena Park, CA 90621

Non Fiction
Natural Disaster Series

Blizzards	1-58659-200-9
Audio Cassette	1-58659-116-9
Audio CD	1-58659-350-1
Earthquakes	1-58659-201-7
Audio Cassette	1-58659-117-7
Audio CD	1-58659-351-X
Hurricanes and Floods	1-58659-202-5
Audio Cassette	1-58659-118-5
Audio CD	1-58659-352-8
Tornadoes	**1-58659-203-3**
Audio Cassette	**1-58659-119-3**
Audio CD	**1-58659-353-6**
Wildfires	1-58659-204-1
Audio Cassette	1-58659-120-7
Audio CD	1-58659-354-4

Cover photo courtesy NOAA Photo Library, NOAA Central Library;
OAR/ERL/National Severe Storms Laboratory (NSSL)
Project Editor: Molly Mraz
Illustrator: Fujiko
Graphic Design: Tony Amaro

ISBN 1-58659-203-3

Contents

Chapter 1

Lillie Matkin was a switchboard operator. She worked in the eight-story R. T. Dennis furniture store in Waco, Texas, in 1953.

May 11 was an uneventful day until the late afternoon. Matkin noticed thunderclouds when she looked out her office window, but she continued working, unconcerned. Suddenly, the lights in her office began blinking on and off. That was more of an annoyance than anything else.

Lillie Matkin did not know what had just happened in San Angelo, Texas. The sky had turned inky black and a funnel-shaped cloud called a tornado dropped from the rolling, tumbling clouds. Eleven people in San Angelo were killed and sixty-six were injured.

A police officer in San Angelo tried to warn the authorities in Waco that the tornado was heading for their town, but the telephone lines blew down before he could get through.

The 90,000 people who lived in Waco had no idea what was about to happen. They were all going about their business as usual. Near downtown, Ira Baden was trimming wood for a garage door. The air was calm but muggy. It seemed unusually dark for 4:30 in the afternoon. Wind gusts started picking up, and then Baden heard a scary sound.

It sounded like several freight trains roaring through Waco at the same time. The clouds seemed to be growling as rain began to fall. The rain was unusual, too. It did not fall straight down to the ground. It moved sideways in a horizontal pattern, pelting the store windows.

Ira Baden looked up and saw the dark clouds swirling violently, like a whirlwind. He knew what was coming--a tornado.

Baden grabbed onto a steel post in the sidewalk and hung on with all his strength. He saw the business district of Waco flying into pieces right before his eyes. The front of a large building was torn away, and he could see panicky people inside. The roof of the Joy Movie Theatre came crashing down.

Baden saw bricks, chunks of cement, lumber, and broken glass whirling around him. The tornado flattened a block of one-story shops. Then it hit the Dennis building, where Lillie Matkin still sat in front of her switchboard. The roof of the Dennis building collapsed in the powerful winds. A funnel cloud hovered right over the building, wrenching the top floors off the ground level. The Dennis building exploded as if filled with dynamite. Dust and debris--broken, scattered remains--sprayed out over a five-block area.

Finally, the tornado raced from Waco, and the residents were left to search for survivors. Ira Baden and many others dug through stones and wood to answer muffled cries for help from trapped people. By 5:00 P.M., the police and fire departments were working furiously to dig out victims. The National Guard joined the effort.

Bright lights were brought in so the rescuers could keep working even after it got dark. A large group of men worked at the mound of debris that had once been the Dennis building. They searched for signs of life. They hacked at the rubble and then called for total silence so everyone could listen for any sounds of survivors. They heard something that might have been a cry for help, but they weren't sure. If there was any chance someone was trapped down there, they would work nonstop to get to that person.

For five hours the men worked, digging a tunnel into the rubble toward the cries they heard. It was hard, painfully slow work that had to be done by hand. Bringing in large equipment could endanger anyone still alive.

So piece by piece they worked. As the tunnel was completed, the men shone their flashlights forward. They heard a woman cry out, "I can see the light! I can see the light!" Now they knew for sure someone was alive in that mass of wreckage, but it would be hours before they could reach the injured woman. By then, it might be too late.

Chapter 2

The trapped woman was Lillie Matkin. The switchboard toppled over when the tornado hit the building, and she was trapped under it.

There was a lot of debris between the rescuers and Matkin. Chunks of cement and wooden beams blocked their path. If they moved a timber the wrong way, the injured woman could be crushed to death.

It took another five hours of careful work before the men reached Matkin. They found her bruised but not seriously hurt. When the switchboard fell, it left her enough room so she wasn't crushed.

Speaking words of encouragement to her, the men worked for another four hours before they got Lillie Matkin to the waiting ambulance. She was imprisoned under the Dennis building for fourteen hours, but she survived.

The tornadoes were on the ground for only a few minutes in San Angelo and Waco, but they wiped out vast areas. Two square miles of Waco's business district was gone. The few modern steel and concrete buildings had withstood the winds quite well. Almost two hundred older buildings were either totally destroyed or so badly damaged that they had to be demolished for safety reasons. In the ruins of Waco, 114 dead were found. Hundreds more were injured.

The funnel cloud that comes with a tornado brings destruction in many ways. The winds batter and twist, and the tornado can create destruction within structures that cause them to

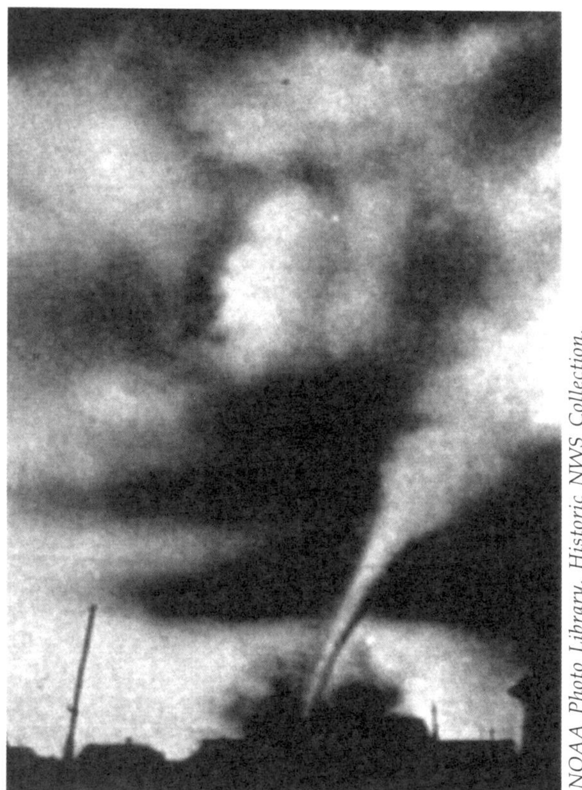

NOAA Photo Library, Historic NWS Collection.

This May 12, 1896 photograph of a Oklahoma City tornado shows the funnel cloud.

fly apart. Tornado winds can lift people and objects high into the air and then hurl them back down to earth.

Tornadoes occur all over the

world, but they happen most often in a part of the United States called Tornado Alley. One-third of all tornadoes in the United States each year take place here. Tornado Alley is a 460-mile-wide strip of land that includes parts of Texas, Oklahoma, Kansas, Nebraska, and Missouri.

In 1951, two years before the San Angelo-Waco tornado, another violent storm struck Texas. Roy Hall and his wife were sitting on their porch in McKinney, Texas. It had been a warm, humid spring, and now the couple watched thunderclouds building up in the west and south.

The clouds turned as dark as tar, and the two systems of clouds merged. Hall saw an amazing curtain of dark green rain and heard a distant roar. The roar grew louder and lightning flashed through the sky. Hailstones as large as baseballs started to crash down on Hall's roof.

Hall and his wife rushed inside the house as tornado winds slammed against the walls. Hall shouted for his wife and children to hide under the bed. He was so fascinated by the awesome show outside that he risked taking one last look before taking shelter with his family.

Hall saw his neighbor's trees being ripped from the ground like weeds. Suddenly, Hall's house shook violently and the room brightened in an eerie way. Everything inside the house was bathed in a strange, bluish light.

Papers and magazines started flying around the room, and Hall was lifted off his feet and thrown ten feet across the floor. The house was torn off its foundation and then was hurled against a group of trees in the front yard.

Most of the roof was gone and one side of the house was caved in. Still, the bluish glow lit up everything. Hall was now lying on his back on the floor of what was once the living room. Since the roof was gone, he could look into the sky. He saw pitch-black clouds with a beautiful white light, like a fluorescent light bulb, inside. Long blue streamers trailed from the clouds. Then came total darkness and the most violent winds of all.

Chapter 3

What remained of the Hall house was pounded into rubble. When daylight came to McKinney, Texas, one hundred of his neighbors were dead, but Roy Hall and his family survived without serious injuries.

Tornadoes are the most violent storms on earth. They have more destructive power than any other storm because winds reach speeds of 300 miles per hour.

In the mid-1880s, the United States Army Signal Corps was in charge of weather reporting. They believed then that tornadoes could be predicted, and warnings could be issued to save lives. But the government of the United States

NOAA Photo Library, Historic NWS Collection.

The Tornado of March 25, 1948 destroyed this large military airplane at Tinker Air Force Base in Oklahoma. Because of that tornado, the U. S. Weather Bureau began to issue tornado warnings.

did not allow such forecasts to be made public because they didn't want to scare people. In fact, the government would not even allow the word *tornado* to be mentioned in weather forecasts. They were afraid that crowds of panicky people trying to leave town would do more damage than the tornado itself. The ban on the word *tornado* finally was lifted in 1938.

Not until March 1952 did the U. S. Weather Bureau issue public tornado warnings. They were not very accurate.

Even now, tornado predicting is far from an exact science. The weather bureau can see a hurricane coming and can track it. But the experts are never quite sure which weather system will produce a tornado.

Tornadoes remain unpredictable. The best that can be done is advising people of the possibility of tornadoes during a period of eight to twelve hours. The warning covers a broad area, and nobody can be sure just when the dreaded funnel cloud might appear. An hour or so in advance of a tornado actually striking, the weather bureau can sometimes issue a warning. If a funnel cloud is spotted, people sometimes have just a few minutes to escape into shelters underground.

"Storm chasers" are people who look for weather conditions likely to produce tornadoes, and then they race down the highways searching for them. The National Weather Service trains the

NOAA Photo Library, NOAA Central Library; OAR/ERL/ National Severe Storms Laboratory (NSSL)

Storm chasers take photographs of tornadoes.

tornado spotters and makes short-term predictions based on their sightings. Storm chasers have taken some of the best photographs of funnel clouds.

One storm chaser saw a whirling cylinder coming down from a cloud. He was not sure whether or not it had actually touched the ground, so he sped on to the next town, where he found serious destruction. He knew then that the tornado he had been chasing already had done its damage.

One of the earliest eyewitness accounts of a tornado striking in the United States comes from New Brunswick, New Jersey, in 1835. On a hot, damp spring evening, dense clouds appeared across the sky. Passengers on a steamer on the Raritan River noticed what some described as a "black and terrible" column dropping from the sky like a funnel.

As it neared the earth, it stirred up dust that rose in a great cloud to meet it. The two clouds joined, looking like a volcanic eruption. Suddenly, the winds were whirling violently. Along the river, trees were yanked from the earth and roofs were torn from houses. There were explosions, and many people smelled sulfur as if something was burning. No fire was found.

After spreading its destruction, the tornado dissolved in a downpour of hail and rain. The frightened people on the steamer lived to tell what they had seen.

Wisconsin County Clerk John M. Bennett gave one of the first detailed eyewitness accounts of a tornado striking. On June 28, 1865, Bennett was driving his horse and buggy home when he noticed dark clouds filling the sky. Summer storms were common to Viroqua, Wisconsin, and Bennett hurried his horse along to avoid getting wet.

There was a strong breeze, and the clouds sped through the sky in a way Bennett had not seen before. He was glad when he got home, put his horse into the stable, and went inside his house.

By 3:00 P.M., it was so muggy that the air felt wet against Bennett's face. By 3:45, there were loud claps of thunder and a threatening roaring sound coming from the sky.

Chapter 4

John Bennett noticed one ragged storm cloud moving swiftly south, and he was fascinated by it. He stood in his garden and watched it. The cloud spun like a child's top, just west of the town of Viroqua.

There was a scoop-shaped basin in Viroqua about a half-mile wide, bordered by craggy hills. A violent wind came down and reached into the basin, ripping a large building apart. Massive timbers were lifted 50 feet into the air before being hurled back onto the rocks.

Bennett ran for his life with his wife, his grown daughters, and his six-year-old grandchild. They had a storm cellar to escape if tornadoes struck, and they tried to get inside it before the full fury of the tornado attacked them.

Bennett's wife was almost crushed by a chimney that fell as she ran, but she dodged it. Bennett did not make it to the storm cellar. As his house blew apart, he was pinned beneath one of the roof timbers. His right leg broken, he lay helpless on the floor as the house dissolved around him.

Bennett was wounded by splinters from the smashed timbers. His eyes and hair filled with dust and plaster from the disintegrating walls. Bennett's horse, in the stable about 50 feet from the house, was yanked from the barn and hurled into the remains of the house. The 1,100-pound animal landed very near where Bennett himself lay. When Bennett turned his head, he could see the horse's head only inches away.

But John Bennett and his family survived the tornado. The whirling winds killed seventeen people in Viroqua, then raced across the prairie toward a school filled with twenty-four children. Five children were killed instantly and another died in the arms of a teacher.

A tornado is basically a funnel-shaped mass with violent, rotating winds that comes down to earth from a thundercloud. It might follow a straight path or zigzag across the land. When it zigzags, one house might be destroyed while the house next door remains untouched.

The word *tornado* comes from the Spanish word *tornar*, meaning to turn. The spinning motion of tornadoes begins slowly. Then, when the spiral tightens, the wind speed increases. It is like a skater who spins more quickly when her arms are close to her body.

Witnesses to tornadoes describe many strange phenomenons, such as odd-shaped clouds, various colors, and sounds. The normal funnel cloud is gray-white and then, as it gathers up debris, it darkens. The distinctive loud roar has often been compared with the roar of freight trains.

There are numerous stories of tornadoes doing amazing things. Supposedly, the winds can pluck the feathers from chickens. But some say this happens because the chicken is so frightened by the winds that it suffers "fright molt," meaning the feathers fall out by themselves.

In 1915, a tornado sent wooden splinters into an iron fire hydrant. In 1896, a tornado drove steel rails in Des Moines, Iowa, 15 feet into the ground.

When tornadoes cross lakes, water can be sucked high into the sky and fall back as heavy rain. In 1886, a tornado struck a section of the Mississippi River and people could see the bottom of the river. All the water was gone, sucked out in seconds and then hurled back from 100 feet in the air. Such events have resulted in fables about frogs and fish raining down. Actually, the tornado sucked up the water and everything in it. When it spewed the water back to earth, it was filled with frogs, toads, and whatever else lived in the river or swamp.

The most deadly tornado recorded anywhere in the world took place in the United States in 1925. Called the "Tri-State Tornado," it tore a 219-mile-wide path from Missouri to Indiana on March 18, 1925. As the tornado gathered strength, many people mistook it for low-lying fog. Others thought the dark clouds were smoke from a railroad. Nobody expected the catastrophe that was about to descend.

Chapter 5

The Tri-State Tornado began its journey above the Arctic Circle near Alaska as just another mass of cold air. The cold, dry Alaskan air gathered into a huge low-pressure system and began to move southwest. Not until March 16 did the system even appear on weather maps.

It was over Calgary, Canada, then, and there was no reason to believe it was anything more than a typical storm. But during the next thirty-six hours, the system sped southward, reaching the Oklahoma-Texas border near Witchita Falls, Texas, by nightfall, March 17. It just looked like some bad weather had arrived.

But something else was coming to meet the visitor from Alaska. A mass of warm, wet air from the Gulf of Mexico was drifting north, and the two systems were about to merge, creating a witch's brew of trouble.

On the morning of March 18, the weather bureau found a long trough of low pressure north of Fort Smith, Arkansas. This trough ran from Lake Erie in the north all the way to southwest Texas. The two weather systems--the cold air from Alaska and the warm air from the south--spun into this trough, creating a tornado. The whirling winds would spawn at least eight tornadoes during the next twelve hours of horror. The deadly Tri-State Tornado was on its way to spread panic and destruction.

The tornado first struck the ground at 1:00 P.M. in Reynolds County, Missouri. In a path of destruction three-fourths of a mile wide, the winds leveled everything, tearing out trees by their roots and shredding buildings. Eleven people died--the first of many victims.

The tornado smashed its way across Missouri, heading for the Mississippi River. It had a forward speed of 60 miles per hour. Normally, tornadoes move along the ground at speeds of 20 to 50 miles per hour, though the whirling winds within them can reach speeds of 300 miles per hour.

Most tornadoes last just a few minutes, but some go on for hours. The Tri-State Tornado would pound a wide area for three and a half hours.

The tornado struck Gorham, Illinois, killing ninety residents and leveling the town. Then it moved on to Murphysboro, moving a mile a minute. It took only a few seconds to level 150 blocks in the residential areas of Murphysboro.

In the wreckage, the mangled bodies of 234 people were found. Hundreds more were injured. Eleven huge steam engines in the railroad yard were flipped onto their sides, and fires broke out in the ravaged buildings. When the fire department tried to fight the fire, they discovered the city's water supply had been destroyed in the storm, and they were helpless against the flames.

A few of the residents of DeSoto, Illinois, actually saw the tornado coming. A brakeman on the Illinois Central train watched a large cloud hanging over the town. There were two blinding flashes of lightning followed by noisy claps of thunder. The sky turned into a seething, boiling mass of clouds whose colors changed constantly. A strange tapering cloud, accompanied by a roaring sound, reached down like an accusing finger into the heart of

DeSoto.

The sky turned so dark it was difficult to see more than a few feet ahead. Then, suddenly, houses exploded as if they had been dynamited.

A mother was in bed with her two-week-old baby. She heard the roaring and knew what was happening, but it was too late to escape. She hugged her baby and hid under the blankets as the house shook around her. The roof timbers groaned as they split apart over her head. The entire roof caved in on the mother and baby. Later, rescuers found the woman and child unhurt. The large beam that fell across the bed allowed just enough space for the pair to escape injury while at the same time preventing more debris from crushing them.

But throughout DeSoto were signs of the disaster that had occurred.

Chapter 6

Nearly 100 bodies were found in the debris that was once the town of DeSoto. Many of the people were literally blown apart in the winds. Arms and legs were torn from bodies. Survivors identified the dead by watches and jewelry on arms that were no longer connected to bodies. Many of the dead had been pierced by knife-like splinters of wood. The splinters were driven into bodies and walls with the force of rifle bullets.

By now the tornado had traveled more than 100 miles over the ground and hundreds were dead. But, instead of losing strength, the Tri-State Tornado grew even stronger.

NOAA Photo Library, NWS Historical Collection.

The Tri-State Tornado hit the De Soto public school at 2:45 P.M. on March 18, 1925.

Near Zeigler, the Illinois Central Railroad bridge was lifted from its concrete pillars and tossed several feet. A wooden church twisted and turned and a frame house ended up facing the opposite direction.

The largest town in the path of the tornado was Frankfort, Illinois. The 20,000 people who lived there received no warning that a tornado was heading straight for them. Many people knew that there were deadly tornadoes around the general area, but nobody knew just where the funnel cloud would come down.

In less than two minutes, more than sixty houses in Frankfort were destroyed and 127 people were killed. Small sticks and splinters had been driven into walls like nails. At the railroad yard, the fierce wind had taken a two-by-four board and driven it into the side of a railroad car.

All through Illinois, the tornado prowled like an angry beast. A large touring car was picked up and hurled 225 feet through the air. A school building was snatched up and set down 150 yards away. Incredibly, the sixteen students inside the building survived the wild ride unhurt. Near Crossville, on the Illinois-Indiana border, straws were picked up and driven into tree trunks. The Tri-State Tornado had killed 606 people in Illinois, and now it moved into Indiana.

The town of Griffin, Indiana, was destroyed, and thirty-four people were killed. Still looking like a black cauldron of terror, the winds blew into Princeton, plunging the town into total darkness. Four men were driving home together when the doors of their car were torn off. The men were pulled from the car and thrown alongside the road, unhurt. But the car in which they were riding was torn to pieces and spread over a wide area.

Now, finally, the tornado was ending its terrible journey. It killed its first victim at 1:00 P.M. in Missouri. The last one died in Princeton, Indiana, at 4:18 P.M. On March 18, 1925, in just one day, 695 people died in three states. There were 219 miles of unbelievable rubble.

Thousands were injured and tens of thousands left homeless. Four small towns were erased from the map.

The Tri-State Tornado left scenes

of destruction that reminded World War I veterans of battlefields. During the tornado, trees were stripped of their leaves and forests were turned into treeless meadows. There were ghastly incidents of flying branches impaling people. Bodies with branches through them were found hanging from trees.

The Tri-State Tornado was unusual in its ferocity and widespread scope. It moved along almost like a hurricane. Six hundred to a thousand tornadoes strike the United States each year. They have occurred in every state, at all times of the year and at all hours, but usually they follow a pattern. Most tornadoes strike in the spring, in the early afternoon or evening. The vast majority of them do little or no serious damage. But anyone who has ever been in a deadly tornado never forgets the experience.

In Greensboro, Kansas, farmer Will Keller was examining the damage a hailstorm had done to his wheat crop on June 22, 1928. As he walked through the broken stalks of wheat, he glanced up to see an umbrella-shaped cloud with a greenish-black base in the southwest. He was so upset looking at his ruined wheat that he paid little attention until he saw three tornadoes hanging like daggers from the cloud.

Chapter 7

One of the tornadoes seemed to be headed straight for the Keller farmhouse, so Keller and his family made a run for the storm cellar. All but Keller himself got safely into the cellar before the storm hit. Keller hesitated so he could get one last look at the tornado.

Two of the tornadoes looked like ropes hanging from the sky, but the third, which was headed right for the house, looked like a funnel with a dome of clouds on top of it. As the cloud moved over the house, the air became very still. Keller smelled gas, and he had difficulty breathing. He looked up into the cloud and saw an opening about 50 to 100 feet wide, lit by flashes of lightning. A hissing sound came from the cloud as it skipped over the Keller farm in a zigzag fashion. The neighboring house and barn were totally destroyed. The people living there did not even make it to their storm cellar. They lay flat on the ground as the winds passed over them. The family's seventeen-year-old daughter had all her clothing ripped off, but she and her family were uninjured.

The Greensboro tornado, though spectacular to watch, did only scattered damage. In Moorhead, Minnesota, a 1931 tornado smashed into the *Empire Builder*, a railroad train heading east at 60 miles per hour. The violent winds collided with the side of the train, knocking five 70-ton cars off the tracks. The 117 passengers on those five cars were thrown 80 feet into a ditch. Amazingly, only one person died and others suffered minor injuries.

In 1932, devastating tornadoes struck in Alabama, taking 268 lives. In 1936, a series of deadly tornadoes struck Mississippi and Georgia, killing more than 450 people. Known as the Tupelo Tornado, the storm did its worst damage in Tupelo, Mississippi, where 216 people died.

A series of tornadoes that reminded people of the Tri-State Tornado of 1925 struck the United States in 1974. A total of 148 tornadoes traveled more than 2,500 miles east of the Mississippi River. They spread death and destruction from Mississippi and Alabama north to Michigan, New York, and Virginia. For sixteen hours, terror reigned through thirteen states. This was the greatest number of tornadoes that ever occurred in the United States at the same time.

Tornadoes are classified according to the Fujita Scale, which measures the intensity of the tornado's winds. A measurement of F-0 describes a tornado with winds from 40 to 72 miles per hour, causing light damage, such as broken tree branches.

In an F-1 tornado, the winds are 73 to 112 miles per hour, and moderate damage is expected, especially to such frail structures as mobile homes.

The F-2 tornado has winds from 113 to 157 miles per hour, and significant damage is expected: trees are uprooted and roofs are torn from homes.

In an F-3 tornado, the winds range from 158 to 206 miles per hour, and severe damage is certain. Trains can be overturned and large objects tossed around.

At the F-4 stage, winds are from 207 to 260 miles per hour, and vast devastation occurs, with loss of life and entire neighborhoods leveled.

The worst category is F-5, with winds greater than 261 miles per hour. Incredible damage occurs, with even steel-reinforced buildings destroyed. Six of the tornadoes that struck in 1974 were F-5 storms.

On Wednesday, April 3, a cold front moving at 35 miles per hour under jet winds moved across Texas. It collided with a mass of low-lying tropical air coming from the Gulf of Mexico. Conditions for this series of tornadoes were very similar to what happened in the Tri-State Tornado.

At 8:42 A.M. on Thursday, weather satellites indicated squall lines. The clouds were enormous. The tops of the clouds reached 60,000 feet into the sky. The weather bureau feared tornadoes might strike, but they were not sure, and they did not know where they would touch down.

Chapter 8

At 1:10 P.M., a small tornado touched down in Morris, Illinois, doing no damage. But it was the beginning of sixteen awful hours.

At 2:00 P.M. in Bradley County, Tennessee, tornadoes struck. Soon they were attacking McLean and Logan counties in Illinois.

At 2:20 P.M., savage winds ripped into a mobile home park, destroying everything and injuring many people. One injured man was waiting for an ambulance when another tornado came, killing him. All through Indiana, Kentucky, Ohio, and Alabama, the tornadoes were fierce.

After the tornado came through

the small town of Guin in northwest Alabama, a state trooper announced that the town did not exist anymore. Nineteen people died and many were injured. All were homeless.

In Jasper, 40 miles away, the city hall and the police station were blown away. In the town of Tanner, one tornado struck at 7:00 P.M., and another at 7:30 P.M. A brick house with six people inside was picked up and carried 250 feet before falling to the ground in a forest. Remarkably, none of the six people was killed or seriously hurt.

In Indiana, seventeen-year-old Karen Scott and five of her friends were riding in a Volkswagon bus when the tornado caught them on a bridge. The winds rammed into the bus, hurling it into the lake 50 feet away. Karen survived, but all of her friends were lost.

In Brandenburg, Kentucky, twenty-nine school children were playing outside their classroom, totally unaware of what was coming. The violent winds swooped up the children, hurling them in all directions. All were killed. For days later, relatives struggled to identify the bodies of their loved ones.

In Sugar Valley, Georgia, nine-year-old Randall Goble was picked up by the tornado and carried 200 yards before being hurled to the ground. He lived to tell the tale, then learned the tragic news that he had become an orphan. His parents and two sisters all died in the family home, which was smashed by the tornado.

The worst-hit city in the deadly series of tornadoes was Xenia, Ohio, with a population of 27,000 people. Three thousand homes and businesses were destroyed in Xenia as a tornado ripped a 3,000-foot-wide path through town.

Fortunately, the twister hit when most of the children had been dismissed from school for the day, or the death toll would have been horrendous.

Not all the students, however, had gone home from school that afternoon. Fifteen members of Xenia High School's drama club had stayed after class to rehearse for an upcoming play.

Eighteen-year-old Ruth Venuti was standing outside the school waiting for her ride home when she noticed a large black cloud hovering in the sky near the school. It was 4:25 P.M. and Ruth felt sure she was witnessing a tornado about to strike. She remembered the students back in the auditorium and ran into the school to warn them.

When Ruth told the drama class what she had just seen, the English teacher and the students did not believe her. They thought she was exaggerating the danger. So they all went outside to see the cloud Ruth was talking about. What they saw when they looked up was a full-blown horror. A violently twisting black column of deadly force was only 200 yards from the school.

Cars parked in front of Xenia High School began to bounce around. The teacher and all the students, including Ruth, raced back into the school building. But they did not go back to the auditorium. They hurried to the center of the building where they believed the structure was strongest. They crouched down in the halls and waited.

Within seconds, the full deadly force of the storm struck the building.

Chapter 9

The sound of the winds smashing into the school was thunderous. As walls and roofs collapsed, there was the sickening sound of crushed wood and concrete colliding. Large pieces of debris fell on the cowering students. They were pelted with dirt, glass, mud, and wood splinters.

Then, abruptly, the wind stopped and there was an eerie silence. No one was seriously injured, although the entire top floor of the school was gone.

A school bus had been hurled onto the auditorium stage where the drama club had been rehearsing when Ruth shouted her alarm. It now lay upside down. The roof of the auditorium had been blown away.

The tornado nightmare continued until the morning of April 5, when the last of the winds ripped through North Carolina. The 1974 killer tornadoes had climbed to the tops of 3,000-foot ridges and had dipped into canyons 1,000 feet deep. Thirty-four people died in Xenia, Ohio, and more than a thousand were injured. In all the tornadoes that ravaged the country that April, 315 people lost their lives and 6,000 were injured. It was the third-deadliest tornado of the twentieth century.

In May 1985, Dave Kostka was umpiring a Little League baseball game in Wheatland, Pennsylvania. When he saw a tornado coming, he ordered all the players and spectators to get into a concrete dugout to wait out the storm. But he believed he had time to take his seven-year-old niece and a ten-year-old neighbor boy to his mother's house. Kostka put the children in his car and drove toward his mother's home.

When the twister appeared in the road, he tried to dodge it. But the car was lifted up and hurled into a ditch.

Kostka pulled the two uninjured children out and took refuge in another ditch. The tornado winds were still howling around them as Kostka covered the children's bodies with his own. The winds yanked Kostka out of the ditch and hurled him against a wall, killing him. But the children survived, as did the players and spectators back at the dugout.

On April 8, 1998, sixty-seven people were praying at the Open Door Church in Birmingham, Alabama. A warning came over the radio that a tornado was heading for the city. The people in the church lay down on the floor, praying as a roar like a freight train swept over them. They heard hissing and crackling as the roof and the walls of the church split apart. The entire church was demolished, but the people inside suffered only minor injuries.

Others in Birmingham that day did not live to tell the tale. The tornado whipped through the city, leaving piles of bricks and shattered lumber. Roofs were hurled into treetops, and a thousand houses were reduced to sticks. Thirty-four people died.

During the same year, on May 30, the 320 citizens of Spencer, South Dakota, were just sitting down to dinner when the winds started blowing hard from the northwest. Spencer was only one square mile of homes and businesses in the middle of vast fields of corn, wheat, and soybeans.

The winds seemed moderate and

nobody was alarmed. But inside the winds was a whirlwind with speeds of 240 miles per hour. The tornado raced through Spencer in a 200-foot-wide path.

As the tornado bore down on the little town, one couple made a desperate run for their storm cellar door. The husband was running to the door, but the wife stopped to look for the family dog. She was crushed in the debris from the collapsing house. Her husband was knocked down and pinned by fallen timbers before he could get to the storm cellar. When the roof of the house was blown off, the man lay on the floor, helpless, as the hail pelted him from the gaping hole above.

Chapter 10

The man in Spencer survived the hail attack during the tornado that lasted only one hundred seconds, but six others, including his wife, died. Not a single business was left standing. Only twelve homes escaped total destruction. At the city bank, the only thing left unscathed was the bank vault, which sat in the middle of a pile of rubble.

The following year, 1999, a violent tornado struck Chickasha, Oklahoma. As a radio announcer warned of a coming tornado, an eleven-year-old boy and his mother were on their way home in the family car. The woman pulled off the road as the winds increased in velocity. She pushed her son into a culvert and lay on top of him to protect him from the storm.

When the winds lifted her up in the air, her boy was holding onto her hand. She told him that she loved him, and that he should let go of her so he would not be dragged away, too. Her body was found several blocks from the scene, but she had saved her son.

In Del City, Oklahoma, during the same series of tornadoes, a man put his wife in the bathtub of their home, covering her with blankets and pillows to shield her from debris. Then he stood at the window holding a pillow to the glass so, if it broke, the shards would not come flying into the house. The tornado sucked him out the window, killing him, but his wife survived.

A wing torn from an airplane in

the Chickasha airport was later found more than 50 miles away in Moore, Oklahoma.

After the 1999 Oklahoma tornadoes had passed, an amazing story was told. A state trooper noticed a rag doll, covered with mud, hanging from the branches of a tree. When he got closer, he realized it was not a doll but a ten-month-old baby girl, ripped from her mother's arms during the storm. The baby was not seriously injured.

On November 11, 2002, the movie theater at Van Wert, Ohio, was filled with sixty children watching a holiday movie. A warning of tornadoes in the area reached the theater manager. He led the children from the auditorium into hallways and bathrooms in the center of the building. With tornado danger in mind, the theater had been built with a cinder block core where people could take refuge.

It was so crowded in one of the girls' bathrooms that the children were literally piled on top of each other. When the tornado struck, two cars were flung into the auditorium where the children had been sitting. Had the children remained there, many would have been killed or seriously injured. As it was, all sixty children survived unhurt.

Across town in Van Wert, a man was driving his large truck when he saw the funnel cloud coming right at him. He watched in horror as the tornado seemed to roll right into the grillwork of his cab. The huge truck spun around like a toy. The truck driver hung on to the steering wheel during the dizzying ride and he was uninjured.

During the November 2002 cluster of tornadoes, fifty tornadoes struck from the South to Ohio. In Mossy Grove, Tennessee, trailer homes were picked up and tossed through the air. Power lines were ripped out as telephone poles went flying. Wrecked cars were piled on top of each other all over towns. The city streets looked like auto-wrecking yards. Thirty-six lives were lost during these tornadoes.

Tornadoes continue to be a threat to life and property in the United States and around the world. Tornadoes killed 500 people in Bangladesh in 1977 and 145 people in eastern India in 1998. Van Wert, Ohio, was especially well prepared to warn residents of tornado danger, and so they were able to save the lives of the children in the theater. Although tornadoes are still unpredictable, a close watch of weather conditions can be a matter of life and death. The timely warning remains the best defense against the deadly winds.

Bibliography

Bluestein, Howard B. *Tornado Alley: Monster Storms of the Great Plains.* New York: Oxford University Press, 1999.

Brown, Billye Walker and Walter R. Brown. *Historical Catastrophes: Hurricanes and Tornadoes.* Reading, Mass.: Addison-Wesley Publishing Co., 1972.

Holford, Ingrid. *The Guinness Book of Weather Facts and Feats, 2d ed.* Enfield, Middlesex: Guinness Superlatives Ltd., 1982.

Myers, Chad, et al. "Deadly Storms Ravage Midwest and Eastern U.S." CNN Wolf Blitzer Reports, 11-11-

2002.

Mackenzie, Dana. "Tornado Rising." *American Scientist*. 7-01-1999, p. 312.

Rosenfeld, Jeffery. *Eye of the Storm: Inside the World's Deadliest Hurricanes, Tornadoes and Blizzards.* New York: Plenom Publishing Co., 1999.

Whipple, A. B. C., and the editors of Time-Life Books. *Planet Earth: Storm.* Alexandria, Va: Time-Life Books, 1982.